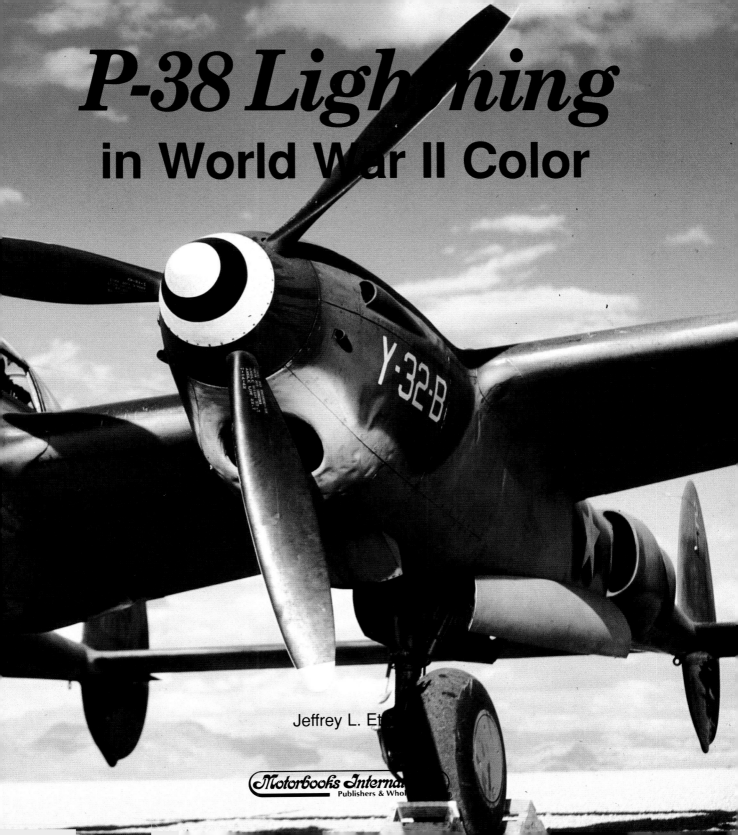

P-38 Lightning
in World War II Color

Jeffrey L. Ethell

Motorbooks International
Publishers & Wholesalers

First published in 1994 by Motorbooks International
Publishers & Wholesalers,
PO Box 2, 729 Prospect Avenue, Osceola, WI 54020 USA

Motorbooks International books are also available at
discounts in bulk quantity for industrial or sales-
promotional use. For details write to Special Sales Manager
at the Publisher's address

Library of Congress Cataloging-in-Publication Data

Ethell, Jeffrey, L.
        P-38 Lightning in World War II color / Jeffrey L.
Ethell
        p.   cm. — (Enthusiast color series)
        Includes index.
        ISBN 0-87938-868-4
        1. Lightning (Fighter planes) 2. World War, 1939-
1945—Aerial operations, American. I. Title. II. Series.
    UG1242.F5E8653   1994
    358.4'383'0973—dc20   93-48650

Printed in Hong Kong

On the front cover: A new P-38E, the first provisionally
combat-equipped model, soars through skies over Florida
during weapons testing at Eglin Field. The E was the first
Lightning to enter combat when used by the 54th Fighter
Squadron in the Aleutian Islands, mid-1942. The large size
of the fighter led Lockheed to incorporate a control wheel
rather than a stick so the pilot could use both hands to get
enough leverage to maneuver the aircraft.
NASM Groenhoff Photo

On the back cover: Two 7th Photo Group F-5s line up for a
formation takeoff at Mt. Farm in late 1944. *Robert Astrella*

On the title page: When the British ordered the Lockheed
Model 322, they named the fighter Lightning, fortunately
ignoring the Lockheed name Atlanta. Unfortunately, the
British ordered their aircraft without turbosuperchargers and
counter rotating propellers, some of the type's major
performance features. This led to a dismal evaluation. The
aircraft already built were given back to the United States
where the Army standardized the P-322 as an unarmed,
advanced Lightning-transition trainer. *USAF*

On the frontispiece: A 1st Fighter Group pilot climbs
aboard his P-38D during the Carolina Maneuvers in late
1941. The fighter is already getting some wear and the
dummy guns are just visible at the front of the nose. Early
Lightnings, through the first Fs, had canopies that opened to
the right side. Later, this was changed, and the hatch opened
toward the rear. The radio antenna wires strung from each
tail came together at a spring, which was then attached to
the rear top of the canopy. The plexiglass stiffeners running
around the rear section of the canopy were replaced by
separating the structure into two sections with a metal brace
running down the center. *NASM Arnold Photo*

Contents

Introduction

Gathering Storm

In the isolationist climate of the 1930s, military aviation budgets suffered severely. Industries improved on older aircraft models rather than build entirely new machines. Yet some visionaries in the services saw the coming world war and realized the United States was ill-prepared. In February 1937, the US Army Air Corps requested proposals for a new type of aircraft, an "interceptor." Congress deemed the idea acceptable because it sounded like a purely defensive weapon, one that would intercept incoming enemy bombers at high altitude.

The Air Corps asked for a major breakthrough in performance: They wanted a minimum true air speed of 360mph at high altitude and the ability to reach 20,000ft in six minutes. Unfortunately, there were no American aircraft engines available with the horsepower to give an airplane such performance figures. The only engine that came close was the Allison V-1710-C8, which had yet to be tested at 1,000hp. At the time,

there wasn't sufficient money for further engine development.

Using the Allison engine, there were two approaches to the problem: build a larger fighter around two of the engines or make an airframe as light as possible using one engine. Fortunately, American manufacturers had an ally in Lt. Benjamin S. Kelsey, the chief of Wright Field's Pursuit Projects Office. Kelsey had coined the term interceptor so the Army could get the fighter it needed without Congress knowing enough to interfere.

Only two proposals were submitted in response to the Army's request—Lockheed's Model 22 (later to become the XP-38) and Bell's Model 4 (later to become the XP-39). Lockheed's Clarence L. "Kelly" Johnson took the more radical approach of designing a twin-boomed, twin-engine fighter that was 150 percent larger than normal. Bell's Bob Woods, on the other hand, put a single engine behind the cockpit, on webs that served as engine mount, wing spar, and landing gear trunion to get a 20 percent reduction in weight. Prototypes were ordered in June and October 1937, respectively.

In July 1938 construction of the XP-38 was initiated using two Allison C-9 engines (later V-1710-11 and -15) rated at 1,090hp each. The right engine turned its propeller "backward" (clockwise when viewed from the front) so the two propellers counter rotated, each engine canceling the effect of the other engine's torque. Maximum weight was projected at over 15,000lb. The new fighter was striking, a vast leap ahead in technology: tricycle landing gear, high wing loading, Fowler flaps for low-speed handling, butt jointed and flush riveted skin, bubble canopy, and metal covered control surfaces.

After the XP-38 was completed in December 1938, it was disassembled, covered, and trucked to March Field, California, during the early morning of January 1, 1939. Kelsey began taxi tests on January 9 and made the first flight on January 27 after delays caused by brake problems and flap flutter. Overall, Kelsey was delighted with the airplane, particularly since it was well within performance requirements. On February 11, after about five hours' total flying time, Kelsey made a speed run to Wright Field, averaging 360mph at cruise power.

Chief of the Army Air Corps, Gen. Henry H. "Hap" Arnold, was there to meet him. He told Kelsey to refuel, fly to Mitchel Field, Long Island, and establish a transcontinental speed record, something he hoped would sell the fighter to Congress. The next leg went flawlessly but, due to the flap flutter and weak-brake problems that had not been ironed out of the prototype, Kelsey had to drag the aircraft in on final approach at minimum speed. Apparently, carburetor ice had formed, leaving the power at idle regardless of throttle movement. Kelsey hit the ground short of the field resulting in the total loss of the XP-38, though he came away unscathed. Arnold took Kelsey to Washington the next day and, fortunately, they sold the program based on Kelsey's reports. Sixty days later, Lockheed had a contract for thirteen service test YP-38s.

Lightning Strike

The first YP flew on September 16, 1940 (the Army had already ordered 673 based on performance estimates) with the last YP being delivered eight months later. By the end of 1941, another 196 aircraft had been delivered, though none were combat capable. Flight testing revealed a problem that was to dog the aircraft through most of its career: compressibility in a dive. Not until dive flaps were added to the J and L models, beginning in June 1944, were pilots allowed to enter prolonged dives. Tail flutter, caused by the sharp junction of wings and fuselage pod, was solved by the installation of large fairings that smoothed the airflow. Though elevator mass balances had been fitted as one of the fixes to this problem, Kelly Johnson insisted they never made any difference.

By the end of October 1941, the first combat equipped version, the P-38E, rolled off the production lines, succeeded by the P-38F in April 1942. The first production aircraft were named Atlanta by Lockheed. Fortunately, the British, who had ordered the type, thought Lightning was a better name and it stuck by October 1941. Unfortunately, the British Model 322 Lightnings were devoid of turbosuperchargers and counter rotating propellers, robbing the fighter of its finest performance features. The Air Ministry canceled its order for 667 Lightnings, but the 143 that had been produced were turned over to the US Army Air Forces for training after being refitted with counter rotating propellers.

Flying the new fighter was a challenge for inexperienced pilots. At full power on takeoff, if one engine failed, full opposite rudder could not keep it from rolling over into the ground. The developed procedure was to reduce power on the good engine, regain control, feather the prop on the dead engine, trim the yaw out, then come back up on the power. Pilots who were not quick enough were killed, resulting in an initial bad reputation for the P-38.

Lightning Storm

With the P-38E, armament was standardized at one 20mm cannon and four .50 caliber machine guns. An additional ninety-nine Es were built as F-4 photographic-reconnaissance aircraft with cameras in place of guns in the nose. These were the first Lightnings to enter combat. In April 1942, they were flown out of Australia, then Port Moresby, New Guinea, with the 8th Photo Reconnaissance Squadron under Maj. Karl Polifka. By June, P-38Es were flying combat missions with the 54th Fighter Squadron up and down the 1,200 mile Aleutian chain in some of the worst weather in the world. From the P-38F production line came F-4As and F-5As; the new F-5 designation remained for all subsequent recce birds.

By November 1942, the P-38F was in action over North Africa with the Twelfth Air Force after being flown across the North Atlantic by the 1st and 14th Fighter Groups, initially attached to the Eighth Air Force. North Africa was a bitter, forsaken theater of operations, hard on both pilots and planes. Seasoned German fighter pilots whittled the Lightnings down as American pilots tried to gain experience as well as prove their new fighters. One of these pilots was the author's father, Lt. E. C. Ethell, 48th Fighter Squadron, 14th Fighter Group. By December, the 3rd Photo Reconnaissance Group, flying F-4s and F-5As, as well as the 82nd Fighter Group, had arrived in North Africa with Lightnings.

In November 1942, the 39th Fighter Squadron took the Lightning into combat over New Guinea, and the 339th Fighter Squadron started flying over the Solomon Islands from the airstrips on Guadalcanal. Evident right away was the inability of the P-38 to outmaneuver the Zero. For the first time, however, superior climb rate, concentrated fire power, and long range put American fighter pilots on the offensive, able to choose when to engage in combat or when to refuse. By December 1942, the 70th Fighter Squadron and the 17th Photo Squadron had arrived at Guadalcanal while the 9th Fighter Squadron was equipped with Lightnings at Port Moresby in January 1943. In March, the 80th Fighter Squadron began operations with the Fifth Air Force out of New Guinea.

On April 18, 1943, a formation of 70th and 339th Squadron P-38s out of Guadalcanal, led by Col. John Mitchell, downed Adm. Isoroku Yamamoto after one of the longest pinpoint interceptions in history. In May 1943, the 475th Fighter Group was created within the Southwest Pacific Theater as the only all-Lightning group. It quickly established a number of scoring records. Throughout the Pacific War, the Lightning proved itself as the premier Army fighter, as both Dick Bong and Tom McGuire proved. Flying only the P-38, Bong became America's ace of aces with forty kills, while McGuire got thirty-eight.

By 1943, the new P-38G was starting to enter service across the globe. With the P-38H came fully automatic supercharger controls for the 1,425hp Allison engines and automatic oil and coolant radiator doors. In the summer of 1943, the 449th and 459th Fighter Squadrons became the only two P-38 fighter units in the China-Burma-India Theater, operating at the end of a long and often forgotten supply line. The 318th

Fighter Group flew the Lightning on long-range missions over the Pacific with the Seventh Air Force.

With victory in North Africa in May 1943, the Twelfth Air Force moved its P-38 groups to Tunisia, then through Sicily and Corsica. In early 1944, they were transferred to the strategic Fifteenth Air Force based at fields around Foggia, Italy. When the last of the P-38Js were being produced in mid-1944, the Lightning had not only dive flaps to prevent compressibility, but hydraulically boosted ailerons, improved cockpit heating, and improved deep chin intercoolers for the turbosuperchargers. Though the early-model P-38s had been more than successful in combat, the new models were the fighter everyone had envisioned, with virtually no limitations. Lightnings in the Mediterranean proved to be extremely effective, and they stayed in combat until the end.

In Europe, the 20th and 55th Fighter Groups took the P-38 into combat in October and November 1943 but lack of cockpit heat and supercharger problems plagued the aircraft well into 1944. By spring 1944, the P-38s of the 370th, 474th, and 367th Fighter Groups had been attached to the Ninth Air Force while the Eighth Air Force got the 7th Photo Group with F-5s. By the end of the war, all fighter units in Europe had transitioned to the P-47 and the P-51, except the 474th, which petitioned Ninth commander Gen. Pete Quesada to let them keep their beloved Lightnings. The final version, the P-38L, had 1,600hp engines. Seventy-five of these were converted into P-38M Night Lightning night fighters but they arrived in the Pacific too late to enter combat. The last of 10,038 Lightnings was rolled out in August 1945.

With the end of World War II, the P-38 Lightning, which had dominated much of the globe, was retired from service. The only reprieve from the smelter came in the form of the revived National Air Races, which lasted from 1946 through 1949, and aerial survey companies who found the F-5 an ideal mount. Lockheed test pilot Tony LeVier and a few others bought surplus P-38s and F-5s for $1,200 or so, then tried to soup them up for racing. Bill Lear, Jr., talked his father into loaning him the money to buy one for fun.

LeVier was by far the most successful Lightning racer, taking second in the 1946 Thompson Trophy Race at 370.1mph, then fifth in 1947. The most thrilling racing event for P-38 lovers was the 1947 Sohio Trophy Race, an all-Lightning event. LeVier took first among seven entrants in his slicked up, all-red P-38L-5.

In combat for just over three and a half years, the P-38 was one of the great American fighters of the war. Only a few remain flying out of the twenty-seven still in existence.

P-38 Lightning Photo Gallery

Elevator mass balances have been retrofitted to this YP-38, a supposed solution to tail flutter. Kelly Johnson was positive they were not needed, particularly since the buffeting was traced to wing-to-fuselage-pod-intersection turbulence, but the company refused to remove them. They became fixtures on every P-38 built, as did the large fairings fitted from pod to wing. Compare this fairingless YP with production Lightnings. *NASM Arnold Photo*

Maj. Signa Gilkey flies one of the thirteen service-test YP-38s. These sleek machines were the first to hit the mysterious phenomenon of compressibility; Gilkey was instrumental in figuring out how to deal with it. When the aircraft hit compressibility in a steep dive, its controls would lock up, making it virtually uncontrollable. No amount of pulling on the control wheel would budge the elevators; the natural reaction for the pilot was to go over the side. Gilkey, however, stayed with a YP in a prolonged dive and found that as it entered denser air he could pull it out using elevator trim. This led to extensive compressibility research and an eventual solution to the problem. *NASM Arnold Photo*

The first YP-38 on the Lockheed ramp at Burbank, September 1940, shows the exceptionally clean lines of Kelly Johnson's creation. Butt joined, flush riveted skin revealed Lockheed's obsession with getting every mile an hour they could out of the design, which had been extensively reworked to improve upon the ill-fated XP-38. *Lockheed Aircraft Corporation*

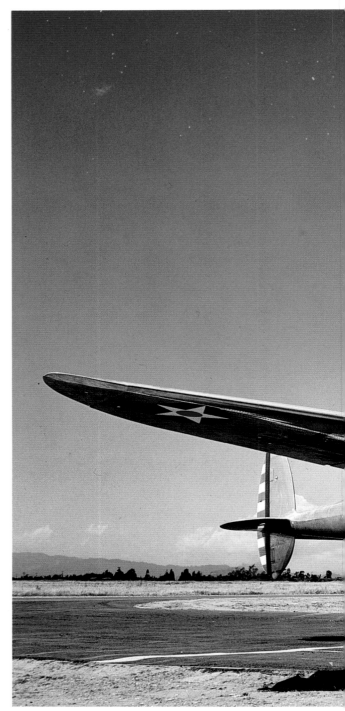

One of the 1st Fighter Group's P-38Ds with dummy guns during the Carolina Maneuvers of November 1941 (the red cross denoted the "enemy" Red Force). Though the early P-38s (there was no letter designation for the first production aircraft) and P-38Ds were not combat capable, they provided Army pilots with invaluable experience in handling this complex and imposing fighter. All of the early P-38s, through much of the E model run, had unpainted Curtiss Electric propellers. *NASM Arnold Photo*

Four P-38Ds on a practice mission, fitted with armament for a change. The early gun installation never incorporated the proposed 37mm Oldsmobile cannon, evident here by the faired-over hole in the center of the nose. Unlike future models, the early Lightnings, if guns were installed, had the four .50 calibers protruding evenly from the nose. Initial training was fraught with peril. If one engine failed on takeoff, the powerful torque of the other engine, no longer countered, could roll the big fighter over before the pilot could react. The exhaust residue from the turbosuperchargers is clearly evident on top of the booms. *NASM Groenhoff Photo*

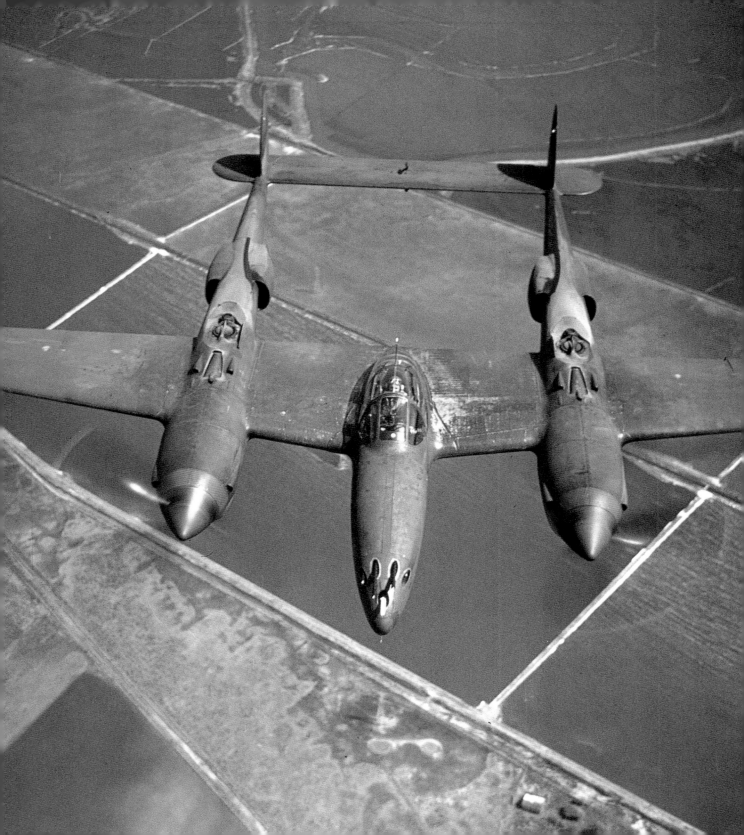

Previous page
A P-38E over southern California irrigated fields during a sortie in mid-1942. The gun installation had now been standardized with staggered .50 caliber machine guns and a 20mm cannon. One of the most common marks of a well-used P-38 was the worn paint on the left wing running up from the trailing edge to the cockpit; this is where the pilot walked after stepping onto the wing from the folding ladder. *NASM Groenhoff Photo*

A new P-38E flies off the southern California coast just after the markings change order of May 12, 1942, which called for the removal of the red circle in the center of the national marking. The red circle still shows through the white star on the boom and the star and blue circle have not yet been repainted on the upper left wing. *NASM Groenhoff Photo*

This P-38E streams oil from its two Allison engines as it climbs for altitude. Another markings change took place in April 1942. Several manufacturers asked for permission to save paint and time in the panic-driven production effort by eliminating "US Army" below the wings of their products. By early May, aircraft were coming off the line without the label, but the marking change was never made official for aircraft already in service, so many retained the lettering. *NASM Arnold Photo*

One of the first batch of P-322-IIs, the last of the British reconversion project, at Shreveport, Louisiana, in January 1943. This non-turbosupercharged Lightning was used by the US Army Air Forces as a high-performance twin-engine trainer, keeping their RAF issue camouflage and serial numbers. *Fred E. Bamberger, Jr.*

The thirtieth P-38F built sits on the line in early 1942. At the time, the arrival of the futuristic P-38 at an Army air base was an event. The Lightning, more than any single Army type at the beginning of World War II, represented what the United States could do technically when the chips were down. The result was a great boost in morale and continual headlines for the P-38 and its pilots. *Fred E. Bamberger, Jr.*

A gunless RP-322 heads out over the Arizona landscape on a transition sortie. Though these "clapped" ex-RAF Lightnings had no high-altitude capability, they could really run on the deck, becoming the near perfect hot rod for new fighter pilots. With no gunnery training until the next phase of fighter lead-in, pilots had nothing to do but learn the fine points of flying the Lightning—in other words, joy ride. *USAF*

A flight of four P-38Es off the southern California coast, early 1942. Clearly, these fighters have seen a great deal of wear. Notice the worn paint from pilots walking up to the cockpit, deep exhaust trails from the turbos, matte paint polished to a sheen from wiping oil and scooting across the wings for maintenance, and oil seeping out at various places. Lockheed photographer Eric Miller took this shot from the open bomb bay of a Hudson. He recalled the Lightnings were being flown by sergeant pilots who later went to the 82nd Fighter Group as flight officers. The lead pilot chomps on a cigar. *Lockheed Aircraft Corp.*

The first fighter Lightnings to enter combat were P-38Es of the 54th Fighter Squadron, which began to fly sorties against the Japanese in the Aleutian Islands in August 1942. This 54th P-38E on jacks undergoes a periodic inspection and repair on the Eleventh Air Force base at Adak, late 1942. The rain, drizzle, cold, fog, and low clouds resulted in far more casualties than the enemy—for both sides. It was a tough theater of war in which to prove the Lightning. *US Army*

A group of 54th Fighter Squadron pilots walk out to their P-38Es before a mission from Adak, Aleutians, in late 1942. Though the clouds are low, the rain has stopped—good weather for this theater of war. At least it was cold all the time and pilots didn't have to worry about sweating in their heavy sheepskin jackets before reaching altitude. Long-range missions over the gray sea, which seemed to melt into the gray sky, were dangerous; pilots often flew into the water when there was no distinct horizon. *National Archives*

Capt. Harry W. Brown in the cockpit of *Sylvia*, his 475th Fighter Group Lightning, New Guinea, late 1943. On December 7, 1941, Brown was one of the few pilots to get airborne over Pearl Harbor in a P-36, downing two Japanese aircraft. He transitioned to Lightnings with the 9th Fighter Squadron, then to the 431st Squadron of the 475th in mid-1943. By the time he returned to the States, he had claimed five more victories. The 475th's P-38s were ridden hard and put away wet. *Dennis Glen Cooper*

A P-38H sits on the line at Williams Field, Arizona, late 1943. With the H model came a number of important improvements: automatic oil and coolant shutters, automatic turbosupercharger controls, and reduced intercooler leakage. The P-38H established the type as the premier fighter in both the Mediterranean and the Pacific theaters, giving pilots such as Tom Lynch, Dick Bong, Tom McGuire, Dixie Sloan, John Mitchell, Bob Westbrook, Danny Roberts, Charles MacDonald, Jerry Johnson, and Jay Robbins an impressive run of kills. *USAF*

As the war entered its third year, older versions of front-line aircraft were relegated to equipment tests and training. This Lightning, flying low over the Florida marshes, has smoke generators mounted on the pylons. The fighter has several different shades of olive drab applied as necessary to keep the metal covered, worn walkways up to the cockpit, and a replacement intercooler leading edge on the left wing. The national star and bar has not been painted back onto the new metal. Fresh olive drab paint has been applied to those areas receiving the most wear. *NASM Groenhoff Photo*

Capt. Willie Haning of the newly formed 475th Fighter Group at Ipswich, near Brisbane, Australia, August 1943. The 475th was the only new group in the Fifth Air Force to form outside the United States in order to save time in getting more P-38s into combat. The Lightning was Fifth Air Force commander Gen. George Kenney's favorite fighter. No one bothered to repaint this P-38 after it had been shipped halfway around the world—the Cosmoline and sealing tape hopelessly marred the pristine factory olive drab and gray paint. *Dennis Glen Cooper*

The first commanding officer of the 475th Fighter Group, Maj. George W. Prentice, stands in front of his P-38 in mid-1943. Prentice brought a wealth of experience to the new Lightning unit, having flown P-40s with the 49th Fighter Group, then commanded the 39th Fighter Squadron when they became the first to fly P-38s in the Southwest Pacific Area (SWPA). Kenney gave him virtual carte blanche to pull old hands from existing units, making the 475th, from the outset, a hot outfit. *Dennis Glen Cooper*

Sitting in a cockpit in the hot Pacific sun could be murderous, leading pilots to remove increasing amounts of clothing until some were flying in nothing more than shorts and G.I. shoes with a parachute strapped on. The extensive plexiglass of the Lightning canopy offered no respite at all, particularly since no part of it could be cracked open for air as in other fighters. If the windows on either side were rolled down, the resulting disturbed air hit the tail, creating an unnerving buffet. *Campbell Archives*

The Headhunters of the 80th Fighter Squadron weren't bragging out of turn with this sign at their headquarters in the Markham Valley of New Guinea when their score stood at 203 victories. The 80th received a total of four Distinguished Unit Citations for combat over Papua from July 1942 to January 1943, New Guinea from August to September 1943, New Britain in October and November 1943, and the Philippine Islands on December 26, 1944. Aces such as Jay T. Robbins (22 kills), George Welch (16 kills, four of which were gained over Pearl Harbor), Ed Cragg (15 kills), Cy Homer (15 kills), and Danny Roberts (15 kills) flew with the 80th. By the end of the war, the 80th had scored 224 of the 8th Fighter Group's 443 victories. *via Bob Rocker/Jack Fellows*

P-38s of the 39th Fighter Squadron and P-39s of the 41st Squadron, both attached to the 35th Fighter Group, sit alert at Dobodura, New Guinea, September 1943. Though the Bell P-39 Airacobra was far outclassed as a fighter, there were never enough Lightnings to replace the aging Bell products and the Curtiss P-40s that continued to fly combat. George Kenney's continual pleas for more P-38s had to compete with commanders in other theaters who wanted them. *via Jack Cook*

When Lockheed built the 5,000th Lightning, a P-38J, they celebrated by painting it a bright fire engine red with the name *Yippee* on the nose and under the wings. Lockheed chief test pilot Milo Burcham flies *Yippee* over the southern California mountains on May 17, 1944. *Eric Miller/Lockheed Aircraft Corp.*

Lockheed chief test pilot Milo Burcham at the controls of the P-38J *Yippee*, the 5,000th Lightning built, May 17, 1944. Burcham was later killed during production testing of the P-80 Shooting Star jet fighter. (The author's father and P-38 gunnery instructor, Capt. Erv Ethell, was one of the pall bearers at Burcham's funeral.) Burcham was considered to be one of the finest test pilots of his era. *Lockheed Aircraft Corp.*

With telltale green spinners and individual aircraft letters on new P-38Js, the 80th Fighter Squadron heads out for takeoff at Nadzab in the Markham Valley of New Guinea. With the J model came virtual mastery of the air over every Japanese stronghold, including the oil fields at Balikpapan, Borneo, that were so vital to enemy fuel production. With aileron boost and improved turbosupercharging, this version of the P-38 was a formidable fighting machine. *via Bob Rocker/Jack Fellows*

An F-5B photo-recce bird is escorted by a P-38J fighter over southern California, fall 1943. The deeper chin under the fighter's engine is the result of moving the intercooler to that location from the wing leading edges, where they never worked consistently. Both aircraft still carry the rounded windshield with armor glass mounted behind inside the cockpit; this was later changed to a flat armor glass windshield. The F-5 carries Synthetic Haze Paint, used as an attempt to hide the airplane in the color of the sky. The scheme consisted of Sky Base Blue with a light shadow shading of Flight Blue, though it takes some real looking to see the differentiation. *Eric Miller/Lockheed Aircraft Corp.*

Capt. Tom McGuire's scoreboard on his P-38J *Pudgy III* at Hollandia, New Guinea. Kills 18 and 19 are freshly painted on, which dates the photo before June 16, 1944, when McGuire downed a Sonia and an Oscar to bring his score to 20. When McGuire was killed in January 1945, he had run his score up to 38 kills, making him second behind Dick Bong. McGuire was a fearless leader with an exceptional ability to fly the P-38. Wingmen said he wrung out his aircraft so hard that they flew "caddywampus" in a yaw after being "sprung." *Dennis Glen Cooper*

When the 318th Fighter Group received P-38s in place of its P-47s for long-range combat, the Lightnings came from several different units outside the Seventh Air Force. This P-38 sits on a coral strip shared by B-29s of the 73rd Bomb Wing, awaiting its next mission. Later P-51s would become the standard escort for Superfortresses hitting Japan. *73rd Bomb Wing Assn. via David W. Menard*

Satan's Angels, the 475th Fighter Group, had some of the most colorful Lightnings in the Pacific. Number 138 flew with the 431st Fighter Squadron, with the unit's Red Devil insignia on the radiator fairing and red trim on everything. The two red stripes on the tail signify the squadron commander, a device used in prewar units, then resurrected by several Pacific fighter groups. By war's end, the 431st had accounted for 225 of the group's 545 victories. *James G. Weir*

This imaginative piece of nose art, *Heady Hedy*, depicts a 44th Squadron, 18th Fighter Group P-38 pilot's dream of Hedy Lamar while stationed at Zamboanga, Mindanao, early 1945. Living in such a remote place, crews found their thoughts turn constantly to girls and food—or food and girls—the order never made much difference. *William Fowkes*

Bill Fowkes with his P-38J *Billy's Filly* at Zamboanga, Mindanao, in the first part of 1945. The 12th Fighter Squadron had moved across the Pacific, flying P-39s from Christmas Island and Efate, then P-38s form Guadalcanal and Treasury Island and on to Sansapor, New Guinea; Morotai; Lingayen, Luzon; San Jose, Mindoro; Palawan; Zamboanga; and Tacloban, Leyte. Flying with any group in the Pacific meant continually pulling up stakes to move to yet another series of tents until the war was finally over. *William Fowkes*

Control tower operators work with the 318th Fighter Group on Ie Shima. Most towers were open on sides, like this one, with a roof that gave only minimal protection from the elements. These guys had to stay at their post—even in driving rain—if aircraft were still airborne and overdue. In many ways, working in the control tower was a thankless job, particularly when more than one shot-up aircraft showed up, each declaring an emergency and asking for priority clearance. *Paul Thomas/Bob Rickard*

Hell's Angel was another 18th Fighter Group P-38J that flew out of Zamboanga in early 1945. The Lightning had an ideal nose for painting all manner of subjects, from dragons to cartoons, some quite elaborate. Often, the artist would remove the gun bay door so he could paint in the relative comfort (well, out of the sun anyway) of his tent. When finished, he'd bring the door back to be rehung. *William Fowkes*

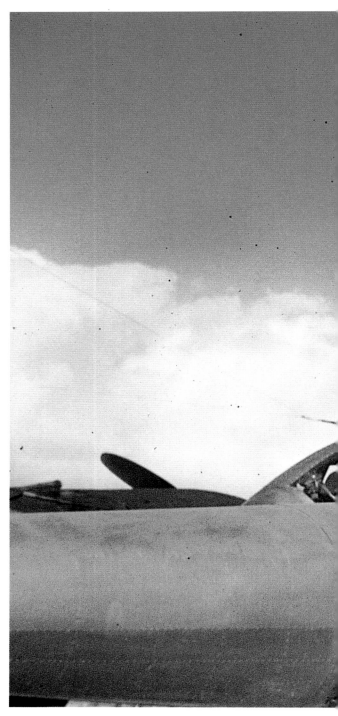

A 318th Fighter Group P-38 on the line at Saipan in December 1944. The 318th received thirty-six well-used P-38s from the 21st Fighter Group in November 1944 and used them well into 1945. The "Lightning Provisional Squadron" did a good job of escort and long-range strike over Iwo Jima, Truk, and other distant targets until newer P-47Ns began arriving in early 1945 to carry on the job. *Paul Thomas/Bob Rickard*

A former 21st Fighter Group Lightning on the line at Saipan. Initially equipped with P-39s as defense for the Hawaiian Islands, the 21st received Lightnings but never used them in combat. In November 1944, thirty-six of their fighters were given to the 318th Fighter Group; in January 1945, the 21st was reequipped with P-51s. The 318th used the P-38s on some long-range escort missions and many came home single engine. *Russ Stauffer via Campbell Archives*

Lt. Howard Byram with his 19th Squadron, 318th Fighter Group P-38 on Saipan. The 318th transitioned from P-47Ds to Lightnings in order to give the Seventh Air Force a deeper striking capability against the Japanese. Pilots found themselves far out across the Pacific on missions that often lasted well beyond seven hours. Over such vast stretches of water, dead reckoning navigation—time and distance—was the only means of finding the way home. On one mission, Byram kept his P-38 going for 9.5 hours, limping back to Saipan from Iwo Jima, before his fuel supply was exhausted. He bailed out with only 25 miles to go and was picked up. The Pacific could be a lonely, forbidding place, particularly when separated and coming home alone. *Howard Byram*

Previous page
A new P-38J-10 cruises over the southern California mountains during a factory acceptance test flight in late 1943. This was the Lightning model that would establish the reputations of many aces and fighter groups, particularly in the Mediterranean and Pacific. Among the improvements for the pilot was a new control wheel with two pistol grips (the same type as on fighters with control sticks) and a flat, bullet-proof windshield. *Lockheed Aircraft Corp.*

At some point in P-38J-10 production from October to December 1943, Lockheed followed the US Army Air Forces directive to eliminate the olive drab and gray camouflage paint in favor of natural aluminum, as this example over California shows. Though pilots worried about their aircraft being more visible, the lack of paint made no difference in combat, other than saving several hundred pounds and possibly giving the pilot a bit more performance. *Eric Miller/Lockheed Aircraft Corp.*

As Jim Weir was undergoing combat transition training in June 1944 at Craig Field, Alabama, before heading for the Pacific, a veteran fighter pilot paid a visit to talk to the pilots and see how their training was going. Upon shutting down, 28 victory ace Dick Bong stepped out of this P-38 and spent some time with Weir and his fellow fledglings. After completing his second combat tour, Bong had been in the States since May 9. *James G. Weir*

Dick Bong taxies out after start-up at Craig Field, Alabama, June 1944, during his stateside visit. Though Bong considered himself "the lousiest shot in the Army Air Forces," he was certainly one of the most talented airplane handlers, able to maneuver up to point-blank range for his kills. Nevertheless, he wanted to learn to shoot properly so he reported to Foster Field, Matagorda, Texas, on July 7, 1944. By the time he graduated on August 6, he had put in over 50 hours of flying time. He returned to the Pacific for his third combat tour and destroyed another twelve Japanese aircraft by December to become America's top ace, with 40 kills. *James G. Weir*

Dick Bong's stateside P-38 at Craig Field, Alabama, during his visit in June 1944 with pilots on their way to the Pacific theater. Though Bong had to put up with a month's worth of bond speeches and publicity, he had come home to take his first serious gunnery training. Bong considered himself a poor marksman. "I am not a good shot," he said in an interview at the time. "I have to hit them either straight from behind or from straight ahead...or with a deflection of not more than ten degrees....I consider it a big accident when I hit anything with deflection shooting." *James G. Weir*

Jim Weir in the cockpit of his 19th Squadron, 318th Fighter Group P-38 on Saipan in late 1944. The 318th's veteran Thunderbolt pilots had racked up quite a record in combat since bringing their big single-engine fighters into combat off aircraft carriers. Transition to the Lightning was met with enthusiasm since it would give the group the range to reach the heavy action. After a few sorties, pilots were, on the whole, pleased to be flying the Lockheed product. *James G. Weir*

318th Fighter Group Lightnings peel off for landing at Saipan, late 1944. The World War II fan break was legendary. An entire squadron would come in toward the field in formation at low level, then break up into the landing pattern with little interval, creating the fan as fighters pitched up in sweeping turns all the way back down to the runway. Gear and flaps would be lowered at about the 180-degree point. The pilot would pull around and roll out just as he passed the end of the runway. Fighter pilots made a contest of seeing who could make the shortest elapsed time from pitch to touchdown. This was later outlawed, however, when pilots pulled too hard and stalled out, often killing themselves.
James G. Weir

Lt. Jim Weir in front of Don Kane's *Killer's Diller*. Though the 318th Fighter Group painted the spinners of each Lightning red, the former 21st Group colors would often come through. In this photograph, the red is peeling off to reveal the original yellow. By February 1945 the Lightnings, which were hand-me-downs from the 21st to begin with, were deemed war weary and ordered transferred to Guam Air Depot. This left the 318th with their worn-out razorback P-47Ds and no worthwhile targets. After a much-resented forced rest, the group was assigned new P-47Ns and reentered combat in late April. *James G. Weir*

A well-dressed Fifteenth Air Force Lightning pilot, early 1944, wears a mixture of American and British equipment: a standard issue A-2 jacket and "pinks" (uniform pants), US Mae West and oxygen mask, and RAF helmet and goggles. The British helmet gave much better ear protection and radio clarity due to a deeper ear cup and better sealing around the ear. Being nonstandard items in the US Army Air Forces, the helmets were high on the list of items for trade. *USAF*

Gunless, all white, *Marjorie Ann*, was a P-38F used as a XV Fighter Command hack aircraft in June 1944. The red spinners were generic for all Fifteenth Air Force fighter groups. Stripped down, the old F was fast and fun to fly, though it finally got so tired and short of spare parts that it was pushed into the scrap heap. *Fred E. Bamberger, Jr.*

Post maintenance engine run, Foggia, sunset, Spring 1944. The right engine on No.13, 48th Squadron, 14th Fighter Group, is just starting to turn over as the crew chief holds the starter button down. The ground crew has pulled an extensive maintenance session in the mud and slime, with frostbitten fingers, and now they have to find out if No. 13 can be released for combat. There was little time to do such things as match the spinner color or even touch up the paint. *Ira Latour*

This Fifteenth Air Force P-38L Pathfinder at Cappodocino Airfield, Naples, June 1945, carries a large radar in the nose for ground mapping and lead bombing through an overcast. The Lightning was ideal for a number of modifications since the nose could hold just about anything. The Pathfinder, one of several Droop Snoot versions that could carry two people, had an external bomb load capacity of 4,000lb. A formation of P-38s dropping bombs aimed by a Droop Snoot could pack quite a wallop. *Fred E. Bamberger, Jr.*

Capt. Warren G. Campbell and his 94th Squadron, 1st Fighter Group P-38J *Old Rusty* at Foggia, Italy. The Lightning carried Campbell through forty-three missions in the Mediterranean Theater of Operations (MTO) where twin-engine reliability was as highly prized as in the Pacific. The 1st was one of the first Lightning groups into combat over Europe and North Africa after having flown across the Atlantic Ocean during Operation Bolero in June 1942. *Warren G. Campbell*

An F-5 of the 3rd Photo Group on short final to its base in Italy. The 3rd moved from England to the Mediterranean just after the invasion of North Africa in November 1942 under the command of the president's son, Col. Elliott Roosevelt, then proceeded to provide vital reconnaissance over Tunisia, Pantelleria, Sardinia, Sicily, Salerno, and Anzio before moving up to Italy and covering the invasion of Southern France. *Peter M. Bowers*

Seen through the Rotol prop blades of a 31st Fighter Group Spitfire VIII, a Free French recce Lightning of the Groupe de Reconnaissance II/33 lands at Pomigliano Airfield, Naples, early 1944. The "Photo Joe" did his job alone and unarmed; when someone failed to return from a mission, no one usually found out what happened. This was the case with the French author Antoine de Saint-Exupéry, lost July 31, 1944, while flying an F-5A with II/33 attached to the 3rd Photo Group. *William J. Skinner*

14th Fighter Group pilot Capt. William Palmer with his *Irene 5th* at Foggia, Italy, in 1944. The 14th, along with the 1st and 82nd Fighter Groups, stayed with their P-38s all through the war. Pilots found the fighter well-suited to combat in the Mediterranean and for long-range escort for the Fifteenth Air Force. Toward the end of the war, all three groups were given an increasing number of air-to-ground attack missions, which were far more dangerous than flying escort. *Norman W. Jackson*

Previous page
A 48th Fighter Squadron, 14th Group Lightning on short final at Triolo, Italy, 1944. The mud and grime on the P-38's wheels and booms is grim testimony to the conditions that dogged Army Air Force units in Italy. Combat did not wait for good weather or lack of mud, so the ground crews worked hard to keep their airplanes ready. *James M. Stitt, Jr.*

This was living, brother—at least at Foggia, Italy, in May 1944. This shack, built out of belly tank crates, belonged to Maj. Doc Crago, the 14th Fighter Group flight surgeon. He was particularly fortunate in having a shade tree to give him some respite from the Italian summer. With little in the way of basic accommodations, Twelfth and Fifteenth Air Force crews used every bit of raw material they could to improve things. *Norman W. Jackson*

In front of *Buzz* are pilot Capt. Norman W. Jackson (left) and crew chief Sgt. S. T. Mamore, 14th Fighter Group, 1944. Jackson had been one of the first replacement pilots to come to the 14th in North Africa in early 1943. He had a long tour of duty before heading back to the States as a P-38 instructor in late 1944. Since all Fifteenth Air Force fighters had red spinners, about the only distinguishing mark 14th Lightnings had was the number on the side of the nose and on the radiator fairing; 131 was assigned to the 49th Fighter Squadron. At times, 14th aircraft had a stripe painted on the tail. *Norman W. Jackson*

The 82nd Fighter Group field at Foggia, Italy, 1945. The yellow-and-red control tower is hard to miss—I guess the tower operators didn't want any flat hatting fighter pilot to run into them. The 82nd was the third P-38 group to enter combat in the Mediterranean, flying their first combat mission on Christmas day 1942. By the time the war ended, the group had claimed 548 aircraft destroyed, another 88 probables, and 227 damaged. *Walter E. Zurney*

Next page
Sgt. Carl Belucher served as crew chief for Lt. Walt Zurney's 97th Squadron, 82nd Fighter Group Lightning *Taffy*. Zurney and Belucher shepherded *Taffy* through fifty combat missions. Zurney ended up flying twelve bombing missions in a Droop Snoot with a glass nose. As he should, Belucher reflects a great deal of pride in his airplane. *Walter E. Zurney*

2nd Lt. Richard E. Gadbury, one of two 97th Fighter Squadron bombardiers, in front of the 82nd Group Droop Snoot loaded for a bombing mission. These bombardiers had a unique job in flying with a fighter outfit since they were rated aircrew but not pilots. The idea was to lead a formation of bomb carrying P-38s with Gadbury or another of his colleagues in the front. All the P-38s dropped their bombs together on Gadbury's "bombs away." *Walter E. Zurney*

An 82nd Fighter Group Lightning with 150gal drop tanks loaded and ready for a mission. Ground crews would start working on their aircraft the minute the pilots shut them down. The crews would go through all the squawks noted by the pilot, rearm the guns, hang new drop tanks, and refuel. By the time the sun went down, the goal was to have the aircraft on the line, ready and in need of only a warm-up the next morning. *Walter E. Zurney*

95th Squadron, 82nd Fighter Group pilots (left to right) Lts. Del Ryland, Larry Peplinski, and Monty Powers, all shack mates at Foggia, in front of a squadron red tail, 1945. The other two squadrons carried different tail colors. The 96th was yellow or white and the 97th was black or blue, resulting in some colorful formations. *Ralph M. Powers, Jr.*

Lt. Monty Powers in the cockpit of his 95th Fighter Squadron Lightning at Foggia, 1945. The Lightning had a roomy cockpit, and pilots found that the control wheel allowed them to put quite a bit of muscle into maneuvering the large fighter. There were few unhappy P-38 pilots, particularly when the J and L models entered combat in 1944. *Ralph M. Powers, Jr.*

Working on a P-38 was a knuckle-busting proposition. The Lightning had two of everything (well, just about), compared to a single-engine fighter, and the cowlings were tight. It was not unusual for mechanics to have skinned, chapped, and bleeding hands. This 37th Squadron, 14th Fighter Group mechanic is already deep into his charge on a sunny afternoon at Triolo. *James M. Stitt, Jr.*

A 48th Fighter Squadron P-38J on takeoff at Triolo, Italy. The most dangerous part of flying the P-38 was the netherworld that existed between liftoff at 90 to 100mph and attaining best single-engine climb speed of 120mph. The pilot's manual was clear: "Be prepared to reduce power immediately to prevent uncontrollable yaw and roll in case of engine failure on takeoff...then apply power gradually and hold enough rudder to prevent the airplane from skidding." Pilots who did not pull the power back to stop the aircraft from rolling over paid for it with their lives. *James M. Stitt, Jr.*

May 1945, the war is over and Venice, Italy, never looked better, particularly from the cockpit of a P-38. While Lt. Glen Supp flew the piggyback fighter around the city, 37th Fighter Squadron engineering officer Capt. Jim Stitt took this shot from his cramped space behind the pilot's seat. With the war over, sight-seeing was the order of the day in every unit with airplanes, particularly for ground crews who had not seen much of the action. *James M. Stitt, Jr.*

A 37th Squadron, 14th Fighter Group crew chief prepares his Lightning at Triolo, Italy, for the day's mission. Though the Mediterranean was supposed to be full of sun and sea, in the winter it was cold and wet, requiring the heavy sheepskin jackets pirated from bomber crews. *James M. Stitt, Jr.*

This happy P-38 pilot, Lieutenant Honeycut of the 37th Fighter Squadron, is wearing his issue khakis instead of a flying suit. In the hot Mediterranean summer, rolled-up sleeves were standard. Though cotton was comfortable, once worn it stayed permanently creased until laundered, particularly in the sweat and heat. *James M. Stitt, Jr.*

In the blowing, swirling dust of Triolo, Italy, 14th Fighter Group Lightnings taxi and take off on a mission in 1944. Sand and grit were major enemies to units flying in the Mediterranean. Since the engine oil was not filtered by anything finer than the screen door mesh in the oil cuno and over the air intakes, dust passed straight into the engine to scour and grind away inside. Engine changes were regular occurrences. *James M. Stitt, Jr.*

Sunset at Triolo, Italy, winter 1944. As engineering officer Jim Stitt remembered, this 37th Fighter Squadron crew chief "will make the late mess—our guys got fed when they finished working and got to the mess hall." Getting the airplanes ready for the next day took precedence over every other activity, including sleep. *James M. Stitt, Jr.*

A 37th Fighter Squadron P-38J warms up in its revetment at Triolo, Italy, winter 1944, before taxiing out to the main runway for takeoff. The Foggia area was flat, ideal for missing obstructions during low approaches but was miserable when the wind picked up or the rain drove across the field or the snow blew through the tents and shacks. *James M. Stitt, Jr.*

A new 55th Fighter Group P-38G at Nuthampstead, England, December 1943. The 55th was the first Lightning unit to enter combat with the Eighth Air Force, flying their first mission on October 15, 1943. The group was also the first to fly over Berlin and by the end of the war, the 55th had destroyed more locomotives in strafing attacks than any other unit. *Air Force Museum*

The 38th Squadron, 55th Fighter Group propeller shop crew works on this P-38J at Nuthampstead. Left to right are T/Sgt. Harold Melby, Cpl. Merle Stivason, Sgt. Robert Sand, and S/Sgt. Kermit Riem. Though Curtiss Electric props were the source of much trouble, particularly in a wet climate like England, these men managed to rack up a record of consistent combat readiness for their P-38s, regardless of the conditions. *Robert T. Sand*

When 338th Squadron, 55th Fighter Group pilot Peter Dempsey went down to strafe a German airfield on May 21, 1944, he ran smack into a set of power lines while trying to fly under them and ended up bringing some wires home. Not only were the rudders locked up, as seen here, but the P-38 was riddled with machine gun fire as well as small- and large-bore flak. The windshield was solid black with oil and one engine was badly damaged, yet Dempsey brought it home to Wormingford. There is little doubt strafing was far more dangerous than air-to-air combat. *Robert T. Sand*

Sunset at Nuthampstead, England, the base of the 55th Fighter Group from September 16, 1943, to April 16, 1944. Personnel well remember the Tannoy loudspeakers that seemed to sprout from most areas of the field. They were extraordinarily clear, particularly when announcing Red Alerts of approaching bombers or—even more nerve racking—of V-1 buzz bombs, nicknamed Doodle Bugs. *Robert T. Sand*

The 55th Fighter Group lining up for takeoff at Wormingford, England, May 31, 1944. The sound of 200 idling engines running through turbos was never to be forgotten. "An all encompassing, but soft rumble" recalls Bob Sand. This day, the 55th was providing fighter escort for the Eighth Air Force heavies attacking marshalling yards and aircraft industry targets in Germany as well as targets in France and Belgium. *Robert T. Sand*

English fog was a constant for Eighth and Ninth Air Forces crews, almost as regular as sunrise. Though weather like this at Nuthampstead on December 26, 1943, would normally ground peacetime operations, the Eighth Air Force often flew, but not this day. The crew chiefs perform maintenance checks, something jumped at on non-mission days. Prop-shop mechanic Bob Sand recalls "sharp memories of this field in dense fog, pitch black night, and trying to tour my guard posts as sergeant of the guard. No headlights of course, and complete disorientation most of the time." *Robert T. Sand*

December 4, 1943. Crew chief Sgt. Red Fraleigh works on Lt. Jerry Brown's 55th Fighter Group Lightning (second from front) at Nuthampstead. Brown made a remarkable trip home single-engine and full of holes after being shot up by a German fighter, which was knocked off his tail by Captain Meyers. Working on a P-38 inside a hangar was a luxury; only the most damaged fighters were put inside. Usually, the crews had to work on them outside. *Robert T. Sand*

Three-quarters of the propeller crew at the line shack (if you can even call it a shack), Nuthampstead, 55th Fighter Group, 1943. Left to right are "Slim" Stivason, Harold Melby, and Kermit Riem. Since there were no permanent facilities on the flight line, crews built and assembled what cover or work spaces they could—at least enough to keep them out of direct rain or something to catch the heat of a potbellied coal stove. *Robert T. Sand*

Pilot Major Hancock and his ground crew, Technical Sergeant Gagnon and Sergeant Witmer, in a huddle before taking off for the first daylight mission to Berlin, "Big B," March 3, 1944. This mission was the one every pilot and crew had been waiting for—and the tension was more than tangible. As it turned out, bad weather forced the bombers back but the 55th Fighter Group didn't get the word and ended up over Berlin alone, the first Eighth Air Force aircraft to fly over the German capital since the war began. *Robert T. Sand*

Malcolm D. "Doc" Hughes' bent-up 7th Photo Group F-5 Lightning at Mt. Farm, England, April 20, 1944. Doc tried to break the low-level buzz job record and ended up flying into the intelligence building, raining white plaster dust on Pappy Doolittle and Walt Hickey, who ran from the building looking like ghosts and thinking they were under attack by a V-1. Doc feathered the left prop, seen here, and nursed his stricken bird to a landing. There were three gashes in the roof that group CO George Lawson ordered immediately repaired to keep Doc out of trouble but a 30-day grounding order was laid down. The group was glad when Doc got back in the air since, out of frustration, he took to driving a jeep like he flew. *Robert Astrella*

Previous page
Shark mouths seemed to have been painted on anything that flew, but the Lightning was ideal due to the shape of its two engine cowlings. This 7th Photo Group F-5C, *The Florida Gator,* went down over France on July 24, 1944, taking Lt. Edward W. Durst with it. Causes of recce aircraft losses were often difficult to ascertain since recce pilots flew alone, but the best guess was that Durst had passed out for lack of oxygen. *Robert Astrella*

Another veteran 7th Photo Group F-5 at Mt. Farm in 1944 shows some use. Though the PRU Blue paint leaps out at the viewer on the ground, in the air it was effective in camouflaging the aircraft. Recon pilots needed every bit of help they could get and the paint did help, but in the end, the pilot's survival depended on skill and a good measure of luck. *Robert Astrella*

This 7th Photo Group F-5 carries invasion stripes on the lower half of booms and wings in the latter half of 1944. Though the stripes were supposed to keep Allied gunners from firing at friendly aircraft, the paint didn't seem to stop it, much to the frustration of photo pilots who often had to drop down low over enemy lines to get their pictures. *Robert Astrella*

Tough Kid, an F-5B, flew a long string of missions with the 7th Photo Group out of Mt. Farm through 1945. The camera ports on the front and side of the nose are evident, but rarely seen is the sighting scope retrofitted in front of the left windscreen. Photo Lightnings were the subject of many modifications, both at the factory and in the field, making each aircraft almost unique. The variety of camera ports and mounts alone was impressive. *Robert Astrella*

Previous page
Two 7th Photo Group F-5s line up for a formation takeoff at Mt. Farm in late 1944. Recce missions were normally single-ship. When a target needed broader coverage, however, two Lightnings would fly spread apart enough for the cameras in both aircraft to photograph a wider area. The fresh paint over repair and reworked spots points out just how much stock military paint would fade—in this case to a lighter blue. *Robert Astrella*

Little Jo, with the 7th Photo Group at Mt. Farm, reflects the official Army Air Forces' elimination of camouflage paint during the last year of the war. Initially, pilots were fearful that a natural metal surface would act like a mirror, reflecting sun to attract enemy pilots or showing up against the dark ground. As it turned out, the weight savings of several hundred pounds and less drag was a plus in combat. *Robert Astrella*

An F-5 taxies out for a mission at Mt. Farm in early 1945. The wet English weather turned the ground into impassable mud. Woe to the pilot who drifted even slightly off the taxiway since even full power would often not be enough to get him out. The pilot would have to get help, usually in the form of the mighty Cletrac "weasel," which seemed able to move just about anything through mud. *Robert Astrella*

Sgt. Julius Cooper, 38th Squadron, 55th Fighter Group painter puts some more paint on a newly arrived P-38J at Nuthampstead, April 11, 1944. When natural-metal aircraft first arrived in England, they were often painted in olive drab and gray camouflage or, if that could not be found, in RAF medium green and gray. These unprimed aircraft often shed their color. Before long, natural metal was accepted as normal and the practice of painting the aircraft at the bases stopped. *Robert T. Sand*

The long perimeter track at King's Cliffe, home of the 20th Fighter Group during its stay in England. The only readily available means of travel around this vast expanse was the ever-present English bicycle—particularly important when wanting to get from the flight line to the mess hall. Group supply records showed ordnance supply procured and issued over 3,000 bicycle parts—and this for only one fighter group; the order must have been enormous for a bomber group. *John W. Phegley*

The dispersal area, 55th Squadron, 20th Fighter Group, King's Cliffe, England, late June 1944. The group still carries the full set of invasion stripes, top and bottom. Drop tanks (150gal) and their crates litter the area behind the P-38J in the foreground, which was lost when 1st Lt. John Klink had to bail out over the English Channel (he was picked up). Though the grass was sectioned off and often reinforced with British wood interlocking plank, it remained a slippery, slimy work area. *John W. Phegley*

A 449th Fighter Squadron P-38J at Chengkung, China, January 1945. Lightnings were rare in the China-Burma-India Theater, so rare that only two individual squadrons were formed, then attached to a line fighter group as a fourth squadron. In the case of the 449th, they were attached initially to the 23rd Fighter Group from August to October 1943, then they were transferred to the 51st Fighter Group through the end of the war. Regardless of what you flew or what you did, this is what good accommodations were like in China.
Fred J. Poats

Looking north at Ontario Army Air Base, just east of Los Angeles, early 1945. (Today, the mountains are still there but smog now obscures this once spectacular view.) Shown is one of the P-38Ls on the line for Lightning transition before the pilots headed for operational outfits. At this point in the war, many combat veterans served as instructors, passing on tips for surviving in hostile skies. *Norman W. Jackson*

Firefighting practice on a wrecked P-38L at Ontario, California, early 1945. With so many junior birdmen at one field trying to master the Lightning, accidents were inevitable, and the base firefighters tried to stay on top of what it took to save a pilot and put out the fire. *Norman W. Jackson*

A P-38L-5-VN, one of the 113 built by Consolidated-Vultee in Nashville, on short final at Ontario Army Air Base, California, in 1945. Not only are the flaps down, but the dive recovery flaps are extended along the bottom of each wing. This simple device allowed Lightning pilots to dive without worrying about compressibility and potential loss of control. *Norman W. Jackson*

Lightnings on the line in Alaska when the war up north was already over. In spite of the Japanese pullout from the Aleutians, the Army Air Forces maintained an air-defense capability in the region until the Japanese surrender. These P-38Ls are in excellent shape in spite of the often terrible weather. *R. Arnold via David W. Menard*

Lightning maintenance at Ontario Army Air Base, California, 1945. Southern California was ideal not only for year-round flying but for ground crews who needed to work on airplanes outside. Though the large hangar could clearly accommodate several aircraft, there were far more on the line around the field than would fit inside. *Norman W. Jackson*

The need for night fighters led to a contract for Lockheed to modify 75 P-38L-5s to P-38M-6 Night Lightnings with AN/APS-4 radar and a second seat for a radar observer. One proof of concept aircraft had already been modified at Hammer Field, Fresno, California, to serve as a prototype. Certainly the overall gloss black paint and red lettering caught the eye. *USAF*

When World War II ended, pilots found themselves in a sudden vacuum. The American war machine had produced almost 300,000 aircraft (among them 10,038 Lightnings), which were now massed all over the globe with nothing for them to do. Fighter pilots begged, borrowed, and stole what flying time they could in the rapid draw down, as Lt. Fred Poats, a 118th Tactical Reconnaissance Squadron P-51 pilot, is doing here at Hangchow, China, in October 1945. Fred gets in a little "exercise" with a 21st Photo Squadron F-5 to keep his skills up while waiting to be shipped back home. Unfortunately, most of the aircraft were so tired and, without wartime pressure to keep them in perfect shape, mechanical emergencies were common. *Fred J. Poats*

Pilot and radar observer climb into a P-38M Night Lightning. The observer had to be small of stature in order to fit into the tiny rear cockpit. The machine guns and 20mm cannon were fitted with anti-flash tubes at the front in order to preserve the pilot's night vision. Though the Night Lightning was shipped to the Pacific in strength, it arrived too late to enter combat. *USAF*

By 1946 the Army Air Forces had an obsolete fleet it didn't know how to get rid of. After all foreign customers had their pick, the P-38s were put up for civilian sale at $1,200 each; the rest were scrapped. These F-5s, P-38Ls, and P-38Ms at Clark Field, Philippines, in February 1946, await their fate under the shadow of Mt. Pinatubo. *via Glenn Horton*

Postwar use of Lightnings centered around two activities: air racing and photo surveying. The unfortunate result of this was the rapid demise of fighter Lightnings in favor of F-5s with good radios and cameras. Nevertheless, those few years of racing (up through 1949) gave the Lightning a colorful swan song. Shown is J. D. Reed's F-5, NX25Y, one of the most colorful racing Lightnings, emblazoned with sponsor Mobil's winged Pegasus. *NASM*

Charles Walling's 25Y had a number of modifications made to streamline it: slimmer fighter-style nose, earlier streamlined cowlings, repositioned carburetor intake, and clipped wings and outer horizontal stabilizers. Unfortunately, it wasn't a real winner against the Corsairs, Mustangs, and Airacobras. Though Walling placed second in the 1947 all-P-38 Sohio Race, he dropped out of the 1947 Thompson due to a balky fuel system. N25Y has survived to the present day, famous as the personal mount of Lefty Gardner. Lockheed test pilot Tony LeVier was by far the most successful Lightning racer, taking second and fifth respectively in the 1946 and 1947 Thompson and first in the 1947 Sohio with his blazing red P-38L-5. The aircraft ended up with Mark Hurd Aerial Surveys where it was destroyed in a crash in 1965. *Aaron King via A. Kevin Grantham*

Spartan Executive became one of the major postwar operators of aerial survey Lightnings, particularly those with Droop Snoot noses for an extra crew member who could operate cameras and line up the photo runs. This one is in particularly good condition as a working airplane, earning its keep and getting the attention it needs. *David W. Menard via Dick Phillips*

By the late 1950s and early 1960s, most warbirds had become so much junk, littering the ramps of airports around the United States. They were valued so low that few people could afford to put money into restoration and the aircraft became too expensive to operate for the return on aerial survey work. Some found their way into museums, like this F-5G at the Tallmantz facility, Orange County Airport, California, June 1966. The aircraft had been Bendix racer No.55 before going to the Honduran Air Force, then to Bob Bean in 1960, then to Tallmantz. David Tallichet ended up with it and put a fighter nose back on it. He flew it for awhile, then traded it to the Air Force Museum, which mounted it on a pole at McGuire Air Force Base, New Jersey, in 1981. *Jeff Ethell*

An Aero Service Corporation F-5G sits with flat tires next to a surplus Corsair in the western United States at about the time it was placed with the Pima County Air Museum by the Air Force Museum. The Aero Service Corporation, a Philadelphia-based company, put Lightnings to use through the end of the type's working career. Age and lack of parts, however, got the best of them and this is what happened. Many were saved, including this one, which was traded to the Musee de L'Air in Paris in May 1989. Unfortunately, it was destroyed a short time later in a disastrous fire that vaporized eighty-five other vintage aircraft. *David W. Menard via Dick Phillips*

Spartan Executive's survey operation spread across North America from Canada to Mexico, a mission particularly suited for the long-ranged Lightning. This Spartan F-5G was photographed from a sister Lightning doing its job over the vast Canadian lake country. Pilots quickly understood the problems faced by wartime bomber escort and photo pilots who sat for long hours in a cramped cockpit at altitude. This time there was only boredom without the momentary terror of combat. *Bob Bolivar via A. Kevin Grantham*

A Droop Snoot Spartan F-5G buzzes one of the Canadian airfields it operated from before the Lightnings were let go. Aerial photography was the last practical use of an airplane that had been designed before World War II. Although operators found it hard to replace as far as operational effectiveness, maintenance and age forced the aircraft out of everyday, on-the-ramp status. *Bob Bolivar via A. Kevin Grantham*

A forlorn Pacific Aerial Surveys, Seattle, two-seat F-5G at Santa Barbara Airport, California, July 1970. This Lightning had flown with Mark Hurd Aerial Surveys from 1953 before being sold to Pacific, who later sold it to Junior Burchinal. It ended up with Merrill Wein who painted it in 459th Fighter Squadron Twin Dragons markings before selling it to Charles Nichols' Yankee Air Corps at Chino. It then was sold to Doug Arnold who had it ferried across the Atlantic, then to Evergreen who brought it back to their facility in Marana, Arizona, where it has been restored by Bill Muszala and his crew. *Jeff Ethell*

Another Pacific Aerial Surveys F-5G at Santa Barbara, July 1970, when all service life had run out on these impressive aircraft. It would be many years before warbird market values climbed enough for buyers to invest in ground-up restorations. The Lightning was perhaps the most complex of wartime fighter aircraft to maintain, not to mention refurbish. *Jeff Ethell*

When Honduras finally let its P-38Ls go to surplus buyer Bob Bean in 1960, there was no great rush for the aircraft. Several sat rotting in the sun, like FAH504 (44-26961), on loan to Ed Maloney's Air Museum, Ontario, California, in October 1967. In 1969 ex-Lightning fighter pilot Larry Blumer purchased the aircraft and painted it in his markings as *Scrapiron IV* after restoration. In 1977 it ended up with John Deahl, who was killed in the crash of the Lightning some time later. The wreck ended up with Lester Friend. *David W. Menard via Dick Phillips*

Though many Lightnings were painted in wartime colors, only Darryl Greenamyer put the red colors of the 5,000th P-38 built, *Yippee*, on his personal airplane, seen here at Van Nuys, California, in June 1966, much the worse for wear. The F-5G had originally been given to Bolivia in the late 1940s under the Military Assistance Plan but never delivered. It was stored in Washington, D.C., until Jack Hardwick bought it in 1950. Ex-14th Fighter Group combat pilot Revis Sirmon bought the aircraft in 1969, built a fighter nose from scratch, and painted it in a fanciful camouflage scheme as N38LL. He flew it continually until a friend of his was killed in the aircraft in 1974. By that time, the warbird movement had picked up enough steam to see the P-38 as a rare find, and to this day it remains the most valuable of all ex-World War II fighters. *Jeff Ethell*

Index